# Table of Contents

# List of Figures

# List of Tables

# List of Abbreviations

| | |
|---|---|
| ADC | Analog-Digital Converter |
| CCD | Charge-Coupled Device |
| CDF | Cumulative Distribution Function |
| DoD | Department of Defense |
| GEODSS | Ground-Based Electro-Optical Deep Space Surveillance |
| LINEAR | Lincoln Near Earth Asteroid Research |
| LRT | Likelihood Ratio Test |
| MIT | Massachusetts Institute of Technology |
| NASA | National Aeronautics and Space Administration |
| NEO | Near Earth Object |
| OTF | Optical Transfer Function |
| PDF | Probability Density Function |
| PSF | Point Spread Function |
| ROC | Receiver Operating Characteristic |
| SNR | Signal to Noise Ratio |
| SST | Space Surveillance Telescope |
| Pan-STARRS | Panoramic Survey Telescope & Rapid Response System |
| USNO | United States Naval Observatory |

# A STUDY OF DIM OBJECT DETECTION FOR THE SPACE SURVEILLANCE TELESCOPE

## I. Introduction

### 1. Problem Statement

Dim object detection is the practice of deciding whether a very dim object, such as an asteroid or faint satellite, is present in a certain location. It, by necessity, deals with trying to determine the presence or absence of low-level signals in the presence of a high system noise. In space surveillance, dim object detection is typically applied to the challenge of locating objects that can be approximated by point sources, such as space debris, satellites, asteroids, and other near earth objects (NEOs).

This research deals primarily with qualitatively and quantitatively comparing the efficacy of two methods for dim object detection, termed in this paper the point detector and the correlator, both of which rely on the likelihood ratio test (LRT). Both methods are a form of Bayesian Analysis, as they rely on the evaluating whether the data collected is the likely result of a pair of hypotheses: one in which an object is present, and the other in which an object is absent. The point detector is based on evaluating the likelihood that a single pixel centered in a window contains an object, and it is currently in use in the Lincoln Near Earth Asteroid Research (LINEAR) and Space Surveillance Telescope (SST) programs. The correlator works by evaluating the likelihood that all the

1

pixels in a window of data result from the propagation of a point source at the center pixel.

## 2. Problem Importance

Dim object detection has multiple applications. It is used in national defense for detecting satellites. It is used to detecting space debris, which threatens both civilian and military assets of both the United States of America and other countries. These are two important tasks whose utility is immediate. Also, it is used in meeting NASA's mandate to detect asteroids [1], which has use both in scientific research and, somewhat less likely, in the possible mitigation of an object from space that would impact Earth.

Obviously, detecting dim objects based on simple inspection is unrealistic; if the objects were easy to distinguish then they wouldn't be dim. For this reason, detecting dim objects is best done with digital signal processing-based methods. Asserting that an object is certainly present is difficult; even more so, asserting that an object is not present on a given trajectory into space is a philosophical conundrum. Thus statistically-based methods that decide the relative likelihood of condition are appropriate.

Given the expense of building, maintaining, and operating large telescopes, it behooves those trying to detect dim objects to implement the best algorithm available, since limited telescope resources are otherwise squandered. To that end, this research is concerned with making the best use of already existing imagery and equipment for the purposes of dim object detection. In essence, employing a better detector is about doing more with less, which is important during this time of fiscal crises. As will be seen, an

2

improved detector does more than just what additional telescope assets would do: it actually allows the detection of objects that would likely not be detectable using optics currently in use. For a given probability of false alarm, a better detector allows for a dimmer object to be detected or for a higher likelihood of detecting an object of given magnitude near the threshold of detection.

### 3. The Space Surveillance Telescope (SST)

This thesis uses imagery of objects in geosynchronous orbit that was taken by the Space Surveillance Telescope (SST). SST was developed to close gaps in current space surveillance methods by doing "broad-area search, detection, and tracking of small objects in deep space[2]." It is able to track space debris, which can be a threat to military and commercial space assets, as well as the assets themselves. It is also capable of detecting other dim objects, such as asteroids, and so it is able to help fulfill NASA's mandate to discover, catalog, and track 90% of all near earth object s (NEOs) that are larger than 140 meters in diameter [1].

It is the first large three-mirror Mersenne-Schmidt telescope. Its 3.5 meter primary mirror, combined with the f/1.0 design allows for an extremely wide field of view that covers an area approximately three times larger than the Ground-Based Electro-Optical Deep Space Surveillance system (GEODSS), which is the best existing geosynchronous earth orbit sensor. As a result it is able to survey a large amount of the sky at once. Its compact design, combined with a highly maneuverable mount and precise steering, allows it to perform a broad search or to track objects as they travel across the sky.

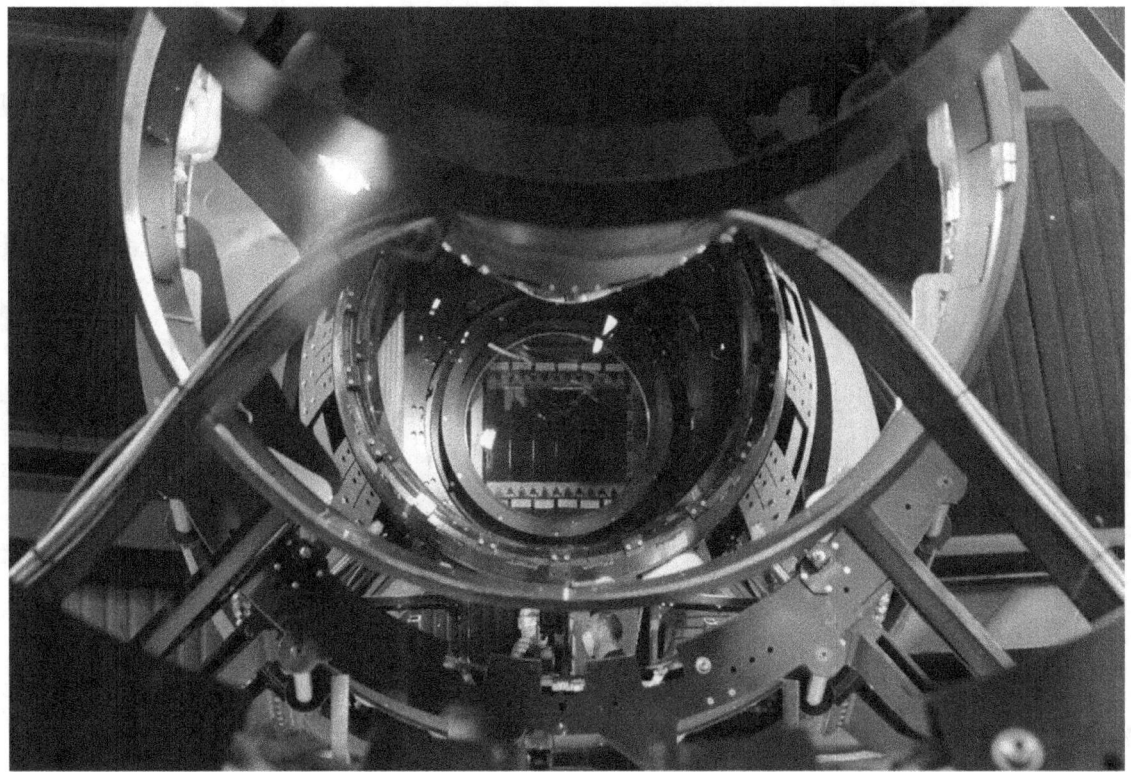

**Figure 1.** View of SST's CCD and wide field of view camera while looking down the optical path. It is easy to distinguish the twelve separate sections of the mosaic focal surface and the supporting electronics that surround it. The telescope is roughly horizontal in this image, so the rim where the walls of the house meet the roof is visible to both the left and right of the telescope. (http://www.darpa.mil/Our_Work/TTO/ Programs/Space_Surveillance_Telescope_(SST).aspx.)

SST employs a unique focal surface charged-coupled device (CCD) that is composed of several pieces of silicon that were knit together into one mosaic and then curved to match the shape of the wavefront. Exposure is controlled with a mechanical, high-speed shutter, rather than controlling the exposure time of the CCD electronically, in order to reduce the complexity of the in-the-chip electronics. Although well-executed, the mechanical limitations of the high-speed shutter prohibit taking images that exhibit a short-exposure optical transfer function (OTF) of the atmosphere, and so all the images

4

that result show the effects of an average, not an instantaneous, atmosphere. All images used in this research were taken while the SST was located at Atom Site in the White Sands Missile Range, above 8000 feet. At the seeing conditions at the Atom Site, the CCD is sensitive enough to be able to detect small objects in geosynchronous orbit, roughly 22,000 miles away.

SST's control and data processing system is able to process the images in near real-time when using the Dirac point detector, which will be described later. The telescope captures up to 1 terabyte of data per night[3]. As will be seen, the research that is presented here can be run on the current data processing system. Should any additional computing resources be need to maintain the ability to run in real-time, then they can be added in parallel, as opposed to extending the depth of the current pipeline.

## 4. Past Research

This section provides a brief review of literature that discusses NEO detection methods and differentiates it from the methods presented in this paper.

### a) Detection without Foreknowledge of the Background

A log-likelihood ratio test used by the Massachusetts Institute of Technology (MIT) Lincoln Laboratory in the Lincoln Near Earth Asteroid Research (LINEAR) program is described in "Applying Electro-Optical Space Surveillance Technology to Asteroid Search and Detection: The Linear Program Results."[4] Some SST data is also processed with this method. That likelihood ratio test (LRT) detector, here-on referred to as the "point detector," is one of two detectors compared in the research effort

5

documented herein. Care has been taken to faithfully duplicate the detector in the manner described in Pohlig's writing, which is discussed in the next section. Particularly, the same threshold of detection that MIT's paper describes, which determines the probability of false alarm, is employed, since this allows a comparison between the proposed correlator and currently implemented system. While LINEAR depends on the presence of an NEO in multiple frames in order to find asteroids, the method being presented in this thesis is strictly single-frame detection.

**b) Detection with Foreknowledge of the Background**

In the January 1989 issue of *IEEE Transaction on Aerospace and Electronic Systems*, "An Algorithm for Detection of Moving Optical Targets" is presented which can be applied to NEO detection[5]. In this paper Stephen Pohlig accurately models the interaction of photons with a charge-coupled device (CCD) as a collection of Poisson random variables [6]. However, in his derivation of the probability of false alarm he departs from a strictly Poisson analysis by using a Gaussian approximation of the normalized log-likelihood function. The method relies on foreknowledge of the background, which does not make it applicable to use with SST in their current processing scheme.

In his 2012 master's thesis *Near Earth Object Detection Using a Poisson Statistical Model for Detection on Images Modeled from the Panoramic Survey Telescope and Rapid Response System*, Capt Curtis Peterson compares the performance of single-frame, log-likelihood ratio test, single point detectors that are based on either Gaussian or

Poisson models photon arrival[7]. He successfully develops a true Poisson log-likelihood detector in which he shows that the threshold of detection can be assumed and then tuned, rather than derived. Capt Peterson's conclusions suggest that further exploration of Poisson-based models will be very fruitful. The methods presented, however, were dependent on the "master sky image" that the Panoramic Survey Telescope & Rapid Response System (Pan-STARRS) generates, which is a running average of the stellar background, and therefore cannot be directly applied to the SST.

## 5. Preview

The remainder of this document will be structured as follows. First, a model of the telescope and the imagery that it is expected to produce is discussed. Then the experiment process and rational is explained. The statistical theory that underlays the both of the detectors is presented before going into the details of how each of the detectors works. The logic by which they are able to be directly compared to each other is then presented. The results of applying the detectors to the experimental data are presented, with an emphasis on comparing the relative probability of detecting dim objects near the threshold of detection using each of the detectors. The computational cost of each of the detectors is then calculated and compared. Conclusions are drawn from the results, and finally limitations of the research are discussed, as well as possible avenues of future work.

## II. The Model and the Experiment

### 1. The Model

A model was created to validate the images taken by the telescope and help explain their characteristics. This model is parameterized to approximate characteristics of the Space Surveillance Telescope (SST), located in the New Mexico desert. It is an F1 telescope with a diameter of 3.5 meters and typical Fried's seeing parameter of 12 cm. The atmosphere is considered to be long exposure and the object to be detected is modeled as a point source.

The transfer function of the telescope can be approximated by propagating a plane wave through a binary screen that represents the aperture of the telescope using a Fraunhofer approximation and the Fourier properties. The approximation is valid since the object is quite distant in relation to the aperture of the telescope such that the distance from any point to the aperture to the object can be considered equidistant. Thus, the transfer function of the telescope , $H_{opt}(u,v)$, is given by Equation (1) [8]

$$H_{opt}(u,v) = \mathcal{F}\left\{ \frac{\left|\mathcal{F}\{p(x,y)\}\right|^2}{\iint \left|\mathcal{F}\{p(x,y)\}\right|^2 dudv} \right\} \tag{1}$$

Here $(x,y)$ are indices that mark off $p(x,y)$, a 1320x1320 binary screen with a circular pupil that is 660 pixels across in the center of the array. $\mathcal{F}$ is the Fourier operator for transforming the operand into $(u,v)$, the frequency space. This binary screen does not account for any obscurations or imperfections in the telescope, but it gives a fairly good approximation of how the telescope itself affects the images, as can be seen in Figure 2.

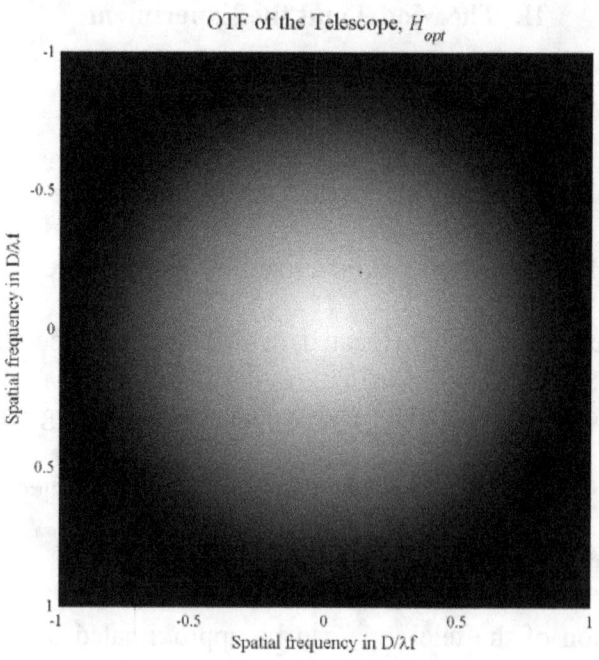

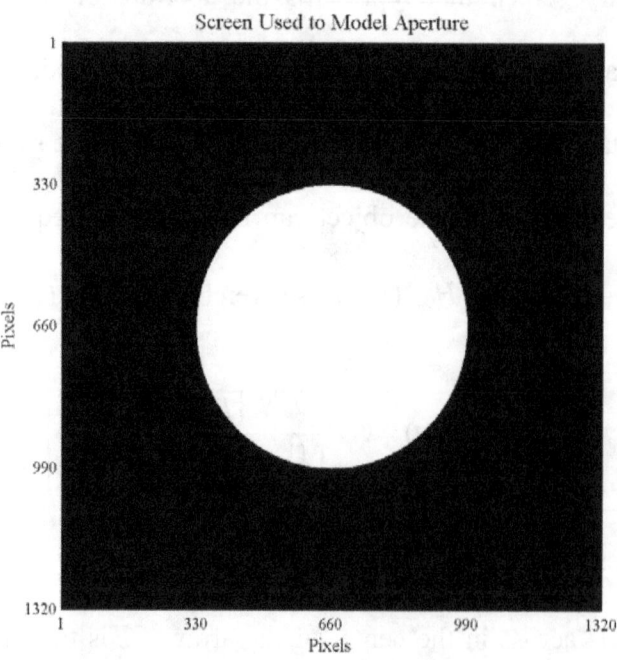

**Figure 2.** Transfer function of the telescope, $H_{opt}$, (top) that results from the propagation of a plane wave through a circular aperture (bottom) using the Fraunhaufer approximation. The intensity of the transfer function denotes the transmissivity of the telescope to different frequencies of light, in terms of $D/\lambda f$.

9

The transfer function of the atmosphere, $H_{atm}(v_{rad})$, is given by Equation (2) [6]

$$H_{atm}(v_{rad}) = \exp\left\{-3.44\left(\frac{\lambda f v_{rad}}{r_o}\right)^{5/3}\right\}$$  (2)

where $\lambda$ is the wavelength of light in micrometers, $f = 3.5\,\text{m}$ is the focal length of the telescope in meters, and $r_o = 12$ cm is Fried's seeing parameter in centimeters (which is typical of the conditions where SST is located). The radial frequency $v_{rad}$ is defined as:

$$v_{rad} = \sqrt{u^2 + v^2}$$  (3)

The resulting pattern is graphed shown in Figure 3.

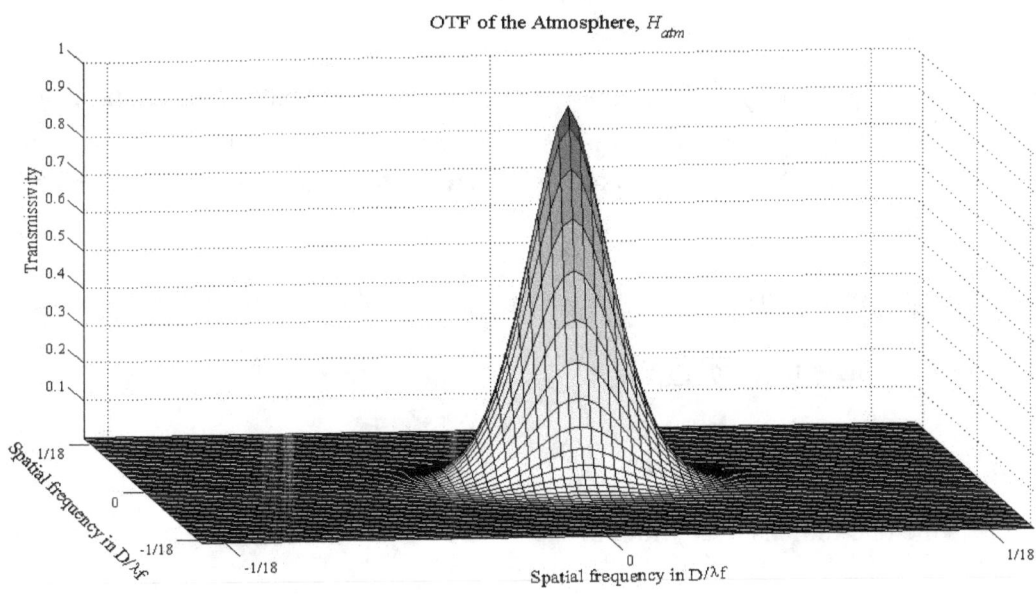

**Figure 3.** Transfer function of the atmosphere, $H_{atm}$, that results from the propagation of a plane wave through the long-exposure atmosphere. The height of the figure denotes the transmissivity of the atmosphere to different frequencies of light, in terms of $D/\lambda f$. Note that the scaling is different from the other transfer functions graphs.

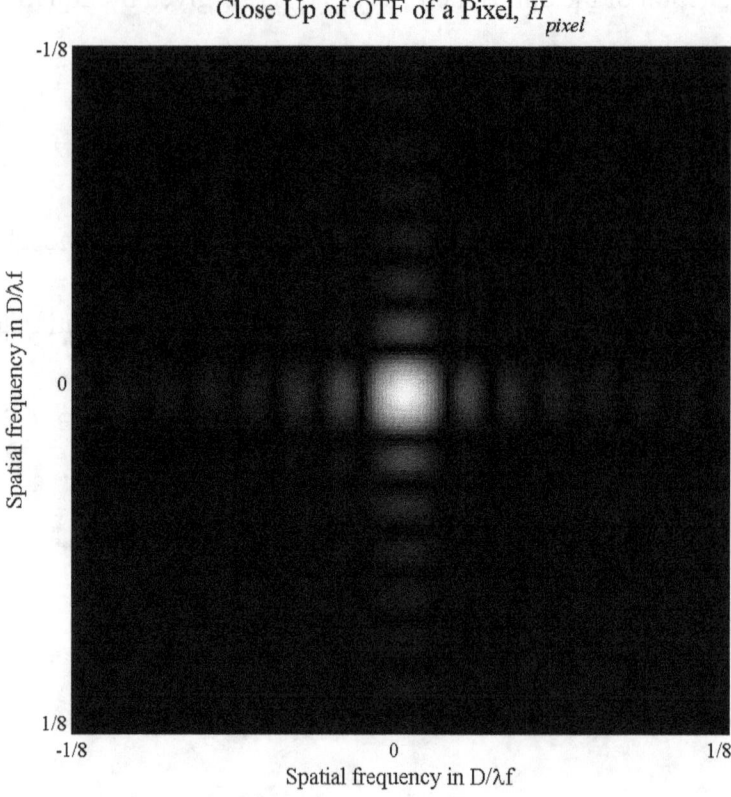

Close Up of OTF of a Pixel, $H_{pixel}$

Spatial frequency in D/λf

Spatial frequency in D/λf

**Figure 4.** Transfer function of the atmosphere, $H_{pixel}$, that results from the propagation of a plane wave through the square aperture of a pixel. The intensity of the figure denotes the transmissivity of the pixel to different frequencies of light, in terms of $D/\lambda f$.

Since each pixel is itself a square aperture, it has the transfer function of a two-dimensional rectangle function [8]:

$$H_{pixel}(u,v) = \mathrm{sinc}(uw)\,\mathrm{sinc}(vw) \tag{4}$$

where $w$ is the one-dimensional width of the square pixel. The resulting pattern is shown in Figure 4.

Since the system is linear, the transfer functions can be combined, as in Equation (5)

$$H_{large}(u,v) = H_{atm}(u,v)H_{opt}(u,v)H_{pixel}(u,v) \tag{5}$$

11

of which the Fourier transform is

$$h_{large}(x,y) = F^{-1}\left\{H_{large}(u,v)\right\} \tag{6}$$

However, the OTF of SST is undersampled by a factor of 30. To properly model this, $h_{large}(x,y)$ is skip-sampled by a factor of 30 to an 11x11 array:

$$\hat{h}(\hat{x},\hat{y}) = h_{large}(x,y), \text{ s.t. } \hat{x} = 30x, \ \hat{y} = 30y \tag{7}$$

and then normalized, so that the PSF of the models is shown in Equation (8):

$$h_{model}(x,y) = \frac{h(\hat{x},\hat{y})}{\displaystyle\iint h(\hat{x},\hat{y})d\hat{x}d\hat{y}} \tag{8}$$

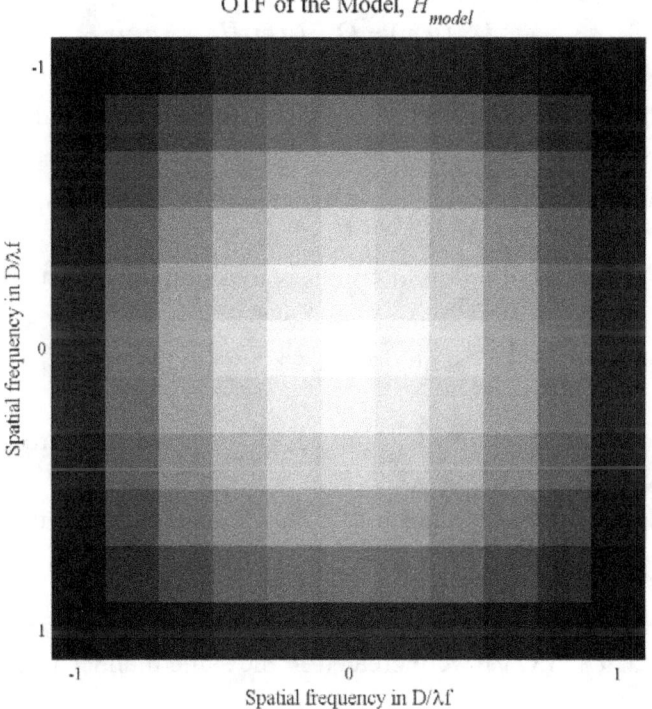

**Figure 5** Transfer function of the model, $H_{model}$, that results from the propagation of a plane wave through the entire system. Note the heavy pixelation, which results from the undersampling of the image plane. The intensity of the figure denotes the transmissivity of the model to different frequencies of light, in terms of $D/\lambda f$.

The OTF of the system, then, shown in Figure 5, is the Fourier transform of the PSF:

$$H_{model}(u,v) = F\{h_{model}(x,y)\} \tag{9}$$

The pristine image of a star, $o_{star}(x,y)$, which is modeled as a point source, is formed by placing a 30 photon-count pixel in the center of a 11x11 zero background. Propagating this image through a telescope or atmosphere is mathematically equivalent with convolving it with the transfer function of a given system, and the result, $i_{star}(x,y)$, is found by [8]:

$$i_{star}(x,y) = o_{star}(x,y) * h_{model}(x,y) \tag{10}$$

which, in the frequency domain, is equivalent to:

$$I_{star}(u,v) = O_{star}(u,v)H_{model}(u,v) \tag{11}$$

Background light and dark current, $N_{dark}(x,y)$, are modeled so that the data, $d_{star}(x,y)$, has a Poisson background with a mean of ten. The measured data is modeled by using this sum as the mean of a Poisson process to account for shot noise:

$$d_{star}(x,y) = \iint o_{star}(w,z)h_{model}(w-x,z-y)dwdz + N_{dark}(x,y) \tag{12}$$

One thousand instances of a noisy star image, $d_{star}(x,y)$, were created in this way to provide a sample of the expected images that SST would produce of dim objects. To provide a sample of a background noise that SST would see, another thousand instances of a noisy background $d_{dark}(x,y)$ were created in the same manner with the same seed to the random number generator, except that where $o_{star}(x,y)$ contained a centered point source on a background of zeros, $o_{dark}(x,y)$ was entirely zeros:

$$d_{dark}(x,y) = N_{dark}(x,y) \tag{13}$$

13

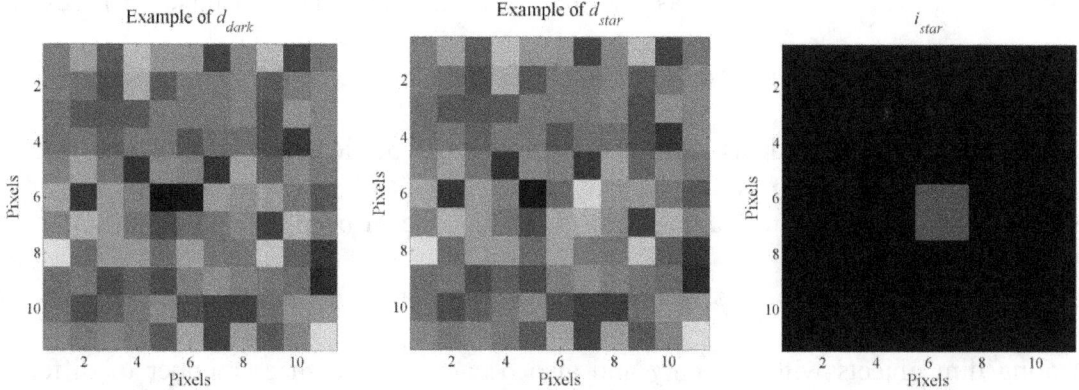

**Figure 6.** Examples from the sample model data sets $d_{dark}$ on the left and $d_{star}$ in the center, both of which have been seeded with the same noise. On the right is $i_{star}$, the result of propagating the point source through the system without adding any noise. The image of $d_{star}$ is only different in that it contains $i_{star}$. The intensity of the point source is three times that of the background. All three images have been scaled equally so that their relative intensities can be judged. Significantly, the brightest pixels are not where the star is located; pixels (1,8) and (11,11) of $d_{star}$ contain the brightest pixels.

Shown in Figure 6 is an example from both sample data sets which has the same noisy background; that is, the only difference is that the example of $d_{star}$ has dim object in the center pixel. As is apparent to the reader, the object is very dim in relation to the background, as will be seen as typical of dim objects in the experimental data near the threshold of detection.

## 2. The Experiment

To perform this research, data was needed of one or more dim objects at varying levels of intensity so that the detection algorithms could be compared. It was necessary to know with certainty the position of the dim object that the detection algorithm was being used to detect so that it was possible to differentiate between false positives and detections, and between the absence of an object and a failure to detect.

Since this research deals with the cutting edge of detection ability and with a telescope that is capable of imaging objects that have never been seen before, using a star catalog to select objects was deemed insufficient, both because of the difficulty of locating dim objects with certainty and also because of the sheer number of different objects that would have to be selected and located in order to characterize the performance of the detectors at different light levels.

Research that studied the glint reflected off satellites during lunar eclipses [9] inspired the experiment, and a glint experiment done by the United States Naval Observatory (USNO) served as the blueprint[10]. In the USNO's experiment, a geostationary satellite was imaged going into eclipse during the autumnal equinox. Data provided by the USNO, which is shown in Table 1 and graphed in Figure 8, shows how a rapid series of images revealed the satellite growing dimmer gradually until it disappears entirely.

In our experiment the recently-constructed SST telescope imaged Anik F1, a geostationary communications satellite, as it went in and out of eclipse for eleven nights between 28 February and 23 March 2012 during the vernal equinox. The satellite, shown in Figure 7, orbits at 107.3° W longitude and is 40.4 x 9 meters at its largest dimensions. The satellite was actively tracked from its orbital elements throughout the entire eclipse. The result is a dataset in which an object of decreasing visual magnitude is at a known location, providing the opportunity to test detection algorithms under a wide variety of light conditions. Since the satellite was tracked, other objects in the sky pass through the

15

**Figure 7.** Artist's depiction of Anik F1 in orbit. (Reproduced with permission from http://www.boeing.com/defense-space/space/bss/factsheets/702/images_702fleet /01pr_01507_hirez.jpg)

frames at the sidereal rate. This situation presents the observer with an object in a known location that has an intensity that varies in time, which allows detection algorithms to be tested over a variety of light levels. These attributes make this type of test ideal for comparing the performance of different detection algorithms.

Data captured once per second, resulting in relatively gradual changes between frames. Both binned (2x2) and un-binned pixel data was collected, but only binned data is included in this analysis. So that a curved imaging surface could be created for SST, it is compromised of several separate pieces of silicon that have been combined together;

**Table 1.** Digital counts for Anik F1 entering the terminator during autumnal equinox. Provided by Dave Monet, USNO

| From USNO Telescope w/Orthogonal Transfer CCD* | | |
|---|---|---|
| *Frame* | *UT* | *Counts above sky* |
| 2500 | 25:31.1 | saturation |
| 2550 | 40:41.1 | 7100 |
| 2560 | 41:31.0 | 2000 |
| 2570 | 42:21.2 | 1100 |
| 2580 | 43:11.2 | 700 |
| 2590 | 44:01.3 | 400 |
| 2600 | 44:51.2 | 200 |
| 2610 | 45:41.2 | 200 |
| 2620 | 46:31.4 | 200 |
| 2630 | 47:21.4 | 150 |
| 2640 | 48:11.2 | 50 |
| 2650 | 49:01.3 | 0 |

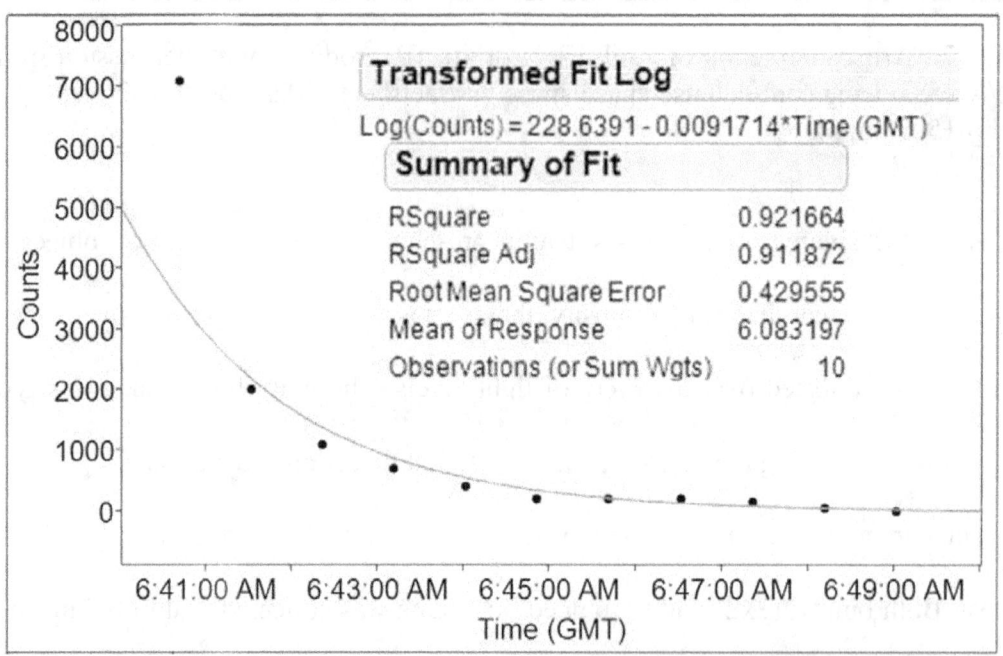

**Figure 8.** Data points and approximate light curve fit of the digital counts for Anik F1 entering the shadow during the autumnal equinox 12 October 2011.

17

typical imaging planes are flat and their silicon is monolithic. As a result, each of the sections of the imaging plane on SST have slightly different responses, and there are visible seams between the pieces. In order to accommodate this, data from only one section of the imaging plane was used in processing a given night.

Data from the nights of March 13-15, 2012 were used in this study, which correspond to nights 073, 074, and 075, respectively. These three nights were chosen because their data is the most comparable to each other. Binned data was collected on these nights. On other nights, either unbinned data was collected or the images suffered from problems like an illuminated moon. On the first two nights the telescope captured the satellite not only as it went into eclipse but as it came out, as well.

The dataset does not provide a basis for forming receiver operating characteristic (ROC) curves. Not only is the amount of data insufficient, it is also extremely difficult to assert that in a given location that an undetectably dim object does not exist, which would be a necessary condition for computing the probability of false alarm. Instead, the probability of false alarm is taken to be fixed and determined by the Gaussian distribution of the log-likelihood detector [4].

### III. Theory and Methodology

### 1. The Likelihood Ratio Test

The likelihood ratio test (LRT) detectors used in this research are binary decision-making processes founded in Bayesian analysis. Use of the LRT for detection of dim objects has been well described before[5][7], and the details of this section deal mainly with their implementation in this particular research. LRTs reach a conclusion by comparing the relative probabilities that the resultant image was the product of two competing hypotheses to a threshold. It is important to understand that the detector cannot respond with absolute certainty that a dim object is or is not present, only which scenario is more likely. The threshold is used to tune the level of certainty with which the detector responds.

The satellite or dim object is considered to be present in a pixel $d(x,y)$ in the image when the likelihood ratio, $\Lambda(x,y)$ exceeds a threshold, $\eta$ , as in:

$$\Lambda(x,y) = \frac{P[d(x,y)|H_1]}{P[d(x,y)|H_0]} \begin{smallmatrix} H_1 \\ > \\ < \\ H_0 \end{smallmatrix} \eta \qquad (14)$$

The two conditional probabilities are the likelihoods that the recorded image resulted from the hypotheses that the satellite is present ($H_1$) or not ($H_0$).

The probabilities are determined based on the statistical characteristics of pixels in an $N \times N$ window of the image that is centered on pixel $(x,y)$. Pixels in this window are referred to by the ordered pair $(w,z)$. For $(x,y)$ to be perfectly centered in the window, it is necessary that $N \in$ odd integers. The upper and lower bounds of the window are defined in relation to $(x,y)$ thus:

$$\begin{aligned} w_{high} &= x + N/2 - 1/2 \\ w_{low} &= x - N/2 + 1/2 \\ z_{high} &= y + N/2 - 1/2 \\ z_{low} &= y - N/2 + 1/2 \end{aligned} \qquad (15)$$

19

Since $N \in$ odd, the addition or subtraction of $1/2$ is necessary to ensure that the bounds refer to integer-numbered pixel. It can be verified that there are precisely $N$ integers in the sets that span $[w_{low}, w_{high}]$ and $[z_{low}, z_{high}]$ when the endpoints are included. These variables are used as the bounds for all summations and products over a window.

As was discussed in the previous chapter, the noise is better characterized as Poisson rather than Gaussian. However, Gaussian distributions well approximate Poisson distributions with means greater than 50, and this research focuses on a comparison using the assumptions that have already been made in the currently fielded point detector. Thus, by assuming that the noise in the image is Gaussian, the probabilities that the object is present ($H_1$) or not ($H_0$) are [7]:

$$P\left[d(x,y)\,|\,H_1\right] = \prod_{w=w_{low}}^{w_{high}} \prod_{z=z_{low}}^{z_{high}} \frac{1}{\sqrt{2\pi}\sigma} e^{\left\{-\frac{1}{2\sigma^2}\left[d(w,z)-B(x,y)-\theta h\right]^2\right\}} \qquad (16)$$

and

$$P\left[d(x,y)\,|\,H_0\right] = \prod_{w=w_{low}}^{w_{high}} \prod_{z=z_{low}}^{z_{high}} \frac{1}{\sqrt{2\pi}\sigma} e^{\left\{-\frac{1}{2\sigma^2}\left[d(w,z)-B(x,y)\right]^2\right\}} \qquad (17)$$

where $w$ and $z$ are the indices that number pixels in a square, $N \times N$ window of the image centered on $d(x,y)$, $B$ is the expected value of the background of the window, $\sigma$ is the standard deviation of the background of the window, $\theta$ is a scale factor for the expected amount of light, and $h$ is the point spread function of the system.

Whereas analysis of data from other experiments like the Panoramic Survey Telescope & Rapid Response System (Pan-STARRS), which is able to make use of foreknowledge of the background, the background, $B(x,y)$, is approximated in this comparison by the mean of all the pixels in the window centered on $(x,y)$:

$$B(x,y) = \sum_{w=w_{low}}^{w_{high}} \sum_{z=z_{low}}^{z_{high}} \frac{d(w,z)}{N^2} \qquad (18)$$

This is a good approximation when the window size is chosen to include the tails of the PSF (point spread function) so that the intensity of most of the pixels in the window is dominated by the background. Should the window be chosen to be too small, then the system will suffer from a lack of contrast between the dim object and the background. Should the window be chosen to be too large, then not only will the computational cost increase unnecessarily, but the risk increases of multiple objects being included in the window. The inclusion of multiple light sources in the window results in a higher computation of the background level, resulting, again, in a decreased contrast and a generally skewed decision-making process. The window used to process all the data in this research was $15 \times 15$ pixels.

The calculation of the standard deviation of the window, $\sigma$, is particular to the implementation of each detector and will be defined below for each detector.

Taking the natural log changes the products to summations, removes the constants, and results in a lower dynamic range for the function [11]. Simplifying, the log-likelihood ratio test, $\mathsf{L}(x,y)$, is:

$$\mathsf{L}(x,y) = \ln \Lambda(x,y) = \sum_{w=w_{low}}^{w_{high}} \sum_{z=z_{low}}^{z_{high}} \frac{(d(w,z)-B)h(w,z)}{\sigma(x,y)} \underset{\underset{H_0}{<}}{\overset{\overset{H_1}{>}}{}} \eta \qquad (19)$$

21

## 2. The Threshold

An examination of the units in Equation (19) shows how the threshold is defined. In the numerator, the pixels in the image, $d(w,z)$, and its mean, $B$, are both in units of photon counts, and their difference represents the signal; its multiplication by the transfer function, $h(w,z)$, which is dimensionless and will be shown to sum to one, does not change the signal fundamentally, but only its representation. In the denominator, the standard deviation, $\sigma(x,y)$, is also in units of photon counts, and it is a measure of the amount of noise in the signal. Thus the threshold, $\eta$, is defined in terms of signal to noise ratio (SNR).

Since for the case of $H_0$ either detector, $\ln \Lambda(x,y)$, will be shown to be a zero mean, unit variance, Gaussian random variable, the threshold is a measure of how likely a false positive is, in terms of standard deviations away from the mean of a Gaussian distribution. This allows the performance of both of the detectors to be compared on the same scale, in particular when $\eta = 6$. For this threshold, an integration of the Gaussian curve shows that the probability of false alarm, $P_{fa}$, is

$$P_{fa} = \int_{\eta=6}^{\infty} \frac{1}{\sqrt{2\pi}} e^{\frac{x^2}{2}} dx = 9.866 e^{-10} \tag{20}$$

## 3. The Point Detector

The "point detector," $L_\delta(x,y)$, is currently in use by LINEAR and SST, and the following explanation of its working is based on Stokes description of it [4]. It is the result of considering only whether the center pixel in each window, $(x,y)$, resulted from

22

propagation of the dim object. This is equivalent to treating the transfer function of the system as a Dirac delta function, $\delta$, a binary screen that blocks out everything except the center pixel, $(x, y)$. Accordingly, the sifting property of the Dirac can be used to make the following simplification for the point detector:

$$\mathsf{L}_\delta(x, y) = \sum_{w=w_{low}}^{w_{high}} \sum_{z=z_{low}}^{z_{high}} \frac{(d(w, z) - B(x, y))h_\delta(w + x, z + y)}{\sigma_\delta(x, y)} \tag{21}$$

$$= \sum_{w=w_{low}}^{w_{high}} \sum_{z=z_{low}}^{z_{high}} \frac{(d(w, z) - B(x, y))\delta(w + x, z + y)}{\sigma_\delta(x, y)} = \frac{(d(x, y) - B(x, y))}{\sigma_\delta(x, y)} \tag{22}$$

For the point detector, $\sigma_\delta$, the standard deviation of the background, is given by:

$$\sigma_\delta(x, y) = \sqrt{\frac{\displaystyle\sum_{w=w_{low}}^{w_{high}} \sum_{z=z_{low}}^{z_{high}} d^2(w, z)}{N^2} - \left(\frac{\displaystyle\sum_{w=w_{low}}^{w_{high}} \sum_{z=z_{low}}^{z_{high}} d(w, z)}{N}\right)^2} \tag{23}$$

$$= \sqrt{\frac{\displaystyle\sum_{w=w_{low}}^{w_{high}} \sum_{z=z_{low}}^{z_{high}} d^2(w, z)}{N^2} - B(x, y)^2} \tag{24}$$

When $H_0$ holds, the structure of the point detector is easily recognized as a prime example of transforming a random variable into standard normal distribution via subtracting by the mean and dividing by the standard deviation.

To verify that $\mathsf{L}_\delta(x, y)$ is zero mean, consider that when an object is not present that the expected value of the center pixel in the window should be the same as the expected value of the window as a whole:

23

$$E\left[(d(x,y)\,|\,H_0\right] = E\left[B(x,y)\,|\,H_0\right] = B(x,y) \tag{25}$$

As a result, the expected value of the numerator is

$$E\left[(d(x,y) - B(x,y))\,|\,H_0\right] = E\left[B(x,y) - B(x,y)\right] = 0 \tag{26}$$

To show that the expected value of the mean is zero, it is sufficient to show that the expected value of the numerator is zero [6]. Thus:

$$E\left[\mathsf{L}_\delta(x,y)\,|\,H_0\right] = \frac{E\left[(d(x,y) - B(x,y))\,|\,H_0\right]}{E\left[\sigma_\delta(x,y)\right]} = \frac{0}{E\left[\sigma_\delta(x,y)\right]} = 0 \tag{27}$$

## 4.  The Correlator

In contrast, the likelihood that all pixels in the window resulted from the propagation of the dim object can be included by using the PSF as the transfer function of the system instead of the Dirac. This detector, $\mathsf{L}_{corr}(x,y)$, termed the "correlator," applies the principles of correlation to general technique used by the point detector. It is based on an LRT that considers the correlation of all the pixels in the window with a point spread function (PSF) of the system, $h_{PSF}$, that is centered on the window:

$$\mathsf{L}_{corr}(x,y) = \sum_{w=w_{low}}^{w_{high}} \sum_{z=z_{low}}^{z_{high}} \frac{(d(w,z) - B(x,y))h_{PSF}(w+x,z+y)}{\sigma_{corr}(x,y)} \tag{28}$$

In this research the PSF was approximated in a simplified manner. For each night, a faint star on the same section of the image plane mosaic was selected as a point source, and its image is used as the basis for the PSF. Several criteria were used in selecting the star so that the resultant PSF was a good approximation of the transfer function of the

24

system. First, the star had to be of an intensity such that the response of the charge-coupled device (CCD) was linear; that is, it had to be bright enough to overcome most of the shot noise but dim enough so that the CCD did not exhibit any saturation or charge overflow. Second, the point source had to be roughly centered on a single pixel; that is, the star did not seem to fall between two pixels or in the corner of a pixel. Third, the bulk of the power had to be confined to roughly a $3 \times 3$ area; in this way, the tails of the PSF were well represented in the $15 \times 15$ window. (Given the simulated transfer functions of the system and the way that stars appear in the images, an object that appeared to be larger could likely not be considered a point source.)

Noise in the raw image of the star, $\overline{d}_{PSF}(x, y)$, is reduced by subtracting $\min(\overline{d}_{PSF}(x, y))$, which is the lowest value found in $\overline{d}_{PSF}(x, y)$, from all of the pixels in $\overline{d}_{PSF}(x, y)$. This removes as much of the background as possible while preserving the tails of the PSF; since the shot noise that dominates the image is strictly additive, doing this subtraction will remove only noise and not any signal. The differenced image, $\overline{d}_{PSF}(x, y) - \min(\overline{d}_{PSF}(x, y))$, is then normalized by dividing by the double summation of

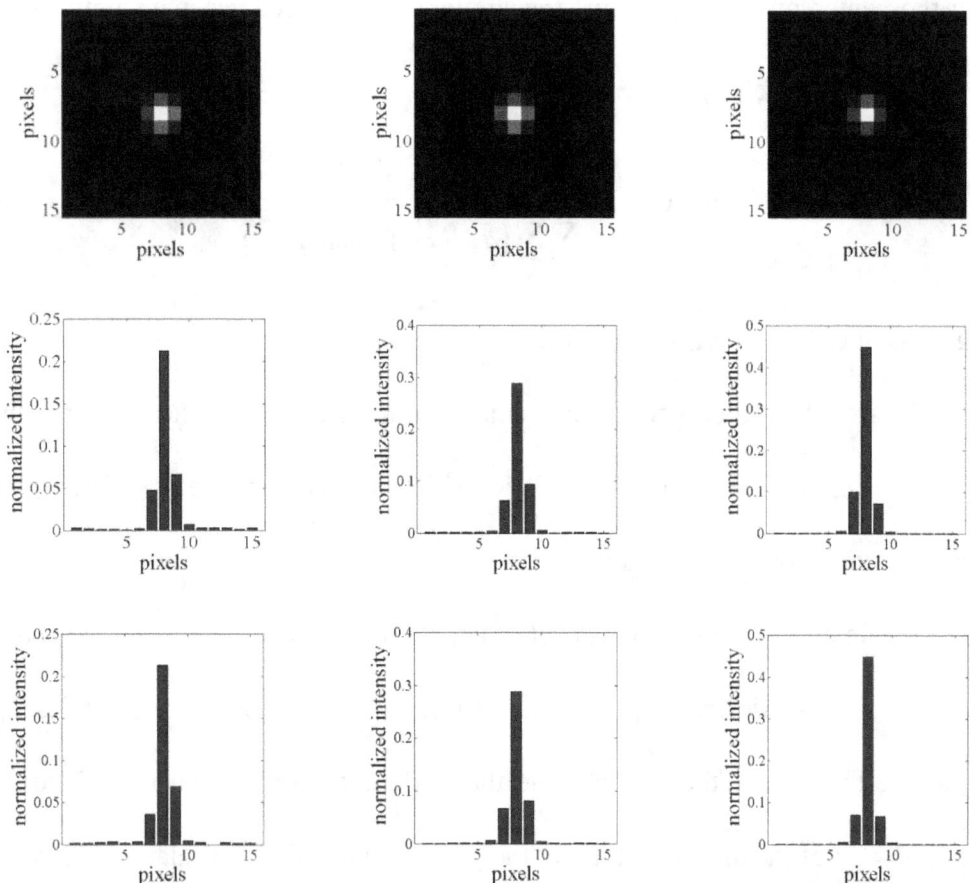

**Figure 9.** The columns show the normalized PSFs used for each of the three nights of data, from left to right: 073, 074, and 075. The first row shows the pixelated image. The second and third rows show the cross sections of the PSFs for the x and y axis, respectively. The cross sections, which are taken across the center row of each axis, show that the bulk of the energy is contained in a $3 \times 3$ square in the center of the window and how the tails roll off towards the edges of the window.

itself, causing the summation over the normalized PSF to equal one; this is done to maintain unity gain in the system so that multiplication by the PSF only changes the shape and does not amplify or attenuate the signal. (Similarly, it can be seen by inspection that the sum over all the pixels of the Dirac function, $h_\delta$, also sums to one since only the center pixel of $h_\delta$ allows transmittance and it does so with unity gain.) Thus, the normalized PSF, $h_{PSF}(x,y)$, is computed by

$$h_{PSF}(x,y) = \frac{\overline{d}_{PSF}(x,y) - \min(\overline{d}_{PSF})}{\displaystyle\sum_x^N \sum_y^N \left( \overline{d}_{PSF}(x,y) - \min(\overline{d}_{PSF}) \right)} \qquad (29)$$

Detailed graphs of each of the PSF's are shown in Figure 9.

These PSFs are taken to be good measurements of the true PSF of the system, rather than estimates. This is due to their high SNR, which results from the stars being very bright compared to the background noise.

A less labor-intensive, more sophisticated approach of determining the PSF could quite possibly yield superior results. There is reason to suspect that since the telescope does not sample the data at the Nyquist rate that differences in the aliasing of the star used as the PSF and of the dim object affect the performance of the correlator. However, as will be seen, the correlator still yields appreciable improvement over the point detector when using a PSF derived from the method described.

As previously discussed, a comparison of the two detectors while using the same threshold necessitates that both detectors need to be zero-mean, unit-variance Gaussian random variables when no dim object is present, as in the $H_0$ hypothesis. To establish these conditions for the correlator, recall that the point detector is defined as

$$\mathsf{L}_\delta(x,y) = \sum_{w=w_{low}}^{w_{high}} \sum_{z=z_{low}}^{z_{high}} \frac{(d(w,z) - B(x,y))\delta(w+x,z+y)}{\sigma_\delta(x,y)} = \frac{(d(x,y) - B(x,y))}{\sigma_\delta(x,y)} \qquad (22)$$

27

and that

$$E\left[\mathsf{L}_\delta(x,y)\,|\,H_0\right]=0 \tag{27}$$

The standard deviation of the correlator, $\sigma_{corr}(x,y)$, will be defined at this point in terms of the multiplication of the standard deviation of the point detector, $\sigma_\delta(x,y)$, and an unknown term, $\sigma_{test}(x,y)$, which will be solved for later:

$$\sigma_{corr}(x,y)=\sigma_\delta(x,y)\sigma_{test}(x,y) \tag{30}$$

The terms for the background and the standard deviation do not change based on the indexing of the variables for the summation. Holding these constant, the definition of the point detector can then be encapsulated with the definition of the correlator:

$$\mathsf{L}_{corr}(x,y)=\sum_{w=w_{low}}^{w_{high}}\sum_{z=z_{low}}^{z_{high}}\frac{(d(w,z)-B(x,y))h_{PSF}(w+x,z+y)}{\sigma_{corr}(x,y)} \tag{31}$$

$$=\sum_{w=w_{low}}^{w_{high}}\sum_{z=z_{low}}^{z_{high}}\frac{\mathsf{L}_\delta(w,z)h_{PSF}(w+x,z+y)}{\sigma_{test}(x,y)} \tag{32}$$

The mean of the correlator can now be determined by substituting in the mean of the point detector:

$$E\left[\mathsf{L}_{corr}(x,y)\,|\,H_0\right]=E\left[\sum_{w=w_{low}}^{w_{high}}\sum_{z=z_{low}}^{z_{high}}\frac{\mathsf{L}_\delta(w,z)h_{PSF}(w+x,z+y)}{\sigma_{test}(x,y)}\,\bigg|\,H_0\right] \tag{33}$$

$$=\sum_{w=w_{low}}^{w_{high}}\sum_{z=z_{low}}^{z_{high}}\frac{E\left[\mathsf{L}_\delta(w,z)\,|\,H_0\right]h_{PSF}(w+x,z+y)}{\sigma_{test}(x,y)}=0 \tag{34}$$

Turning attention to finding the variance, the second moment of the correlator is calculated:

$$E\left[L_{corr}^{2}(x,y) \mid H_{0}\right] = E\left[\sum_{w_{1}=w_{low}}^{w_{hig}}\sum_{z_{1}=z_{low}}^{z_{high}}\frac{L_{\delta}(w_{1},z_{1})h_{PSF}(w_{1}+x,z_{1}+y)}{\sigma_{test}(x,y)}\right.$$
$$\left.\times \sum_{w_{2}=w_{low}}^{w_{high}}\sum_{z_{2}=z_{low}}^{z_{high}}\frac{L_{\delta}(w_{2},z_{2})h_{PSF}(w_{2}+x,z_{2}+y)}{\sigma_{test}(x,y)} \mid H_{0}\right] \tag{35}$$

$$= E\left[\sum_{w_{1}=w_{low}}^{w_{high}}\sum_{z_{1}=z_{low}}^{z_{high}}\sum_{w_{2}=w_{low}}^{w_{high}}\sum_{z_{2}=z_{low}}^{z_{high}}L_{\delta}(w_{1},z_{1})L_{\delta}(w_{2},z_{2})\right.$$
$$\left.\times \frac{h_{PSF}(w_{1}+x,z_{1}+y)h_{PSF}(w_{2}+x,z_{2}+y)}{\sigma_{test}^{2}(x,y)} \mid H_{0}\right] \tag{36}$$

When $(w_{1},z_{1})=(w_{2},z_{2})$, substituting one for the other reduces this to

$$E\left[L_{corr}^{2}(x,y) \mid H_{0}\right] = \sum_{w_{1}=w_{low}}^{w_{high}}\sum_{z_{1}=z_{low}}^{z_{high}}\frac{E\left[L_{\delta}^{2}(w_{1},z_{1}) \mid H_{0}\right]h_{PSF}^{2}(w_{1}+x,z_{1}+y)}{\sigma_{test}^{2}(x,y)} \tag{37}$$

In his article, Stokes established that the point detector is unit-variance, so we can

substitute in $E\left[L_{\delta}^{2}(w_{1},z_{1}) \mid H_{0}\right]=1$ [4]. Thus:

$$E\left[L_{corr}^{2}(x,y) \mid H_{0}\right] = \sum_{w_{1}=w_{low}}^{w_{high}}\sum_{z_{1}=z_{low}}^{z_{high}}\frac{h_{PSF}^{2}(w_{1}+x,z_{1}+y)}{\sigma_{test}^{2}(x,y)} \tag{38}$$

Since we desire for $L_{corr}(x,y)$ to have a unit variance, we define it as such and solve

for $\sigma_{test}(x,y)$:

$$1 = \sum_{w_{1}=w_{low}}^{w_{high}}\sum_{z_{1}=z_{low}}^{z_{high}}\frac{h_{PSF}^{2}(w_{1}+x,z_{1}+y)}{\sigma_{test}^{2}(x,y)} \tag{39}$$

$$\sigma_{test}^{2}(x,y) = \sum_{w_{1}=w_{low}}^{w_{high}}\sum_{z_{1}=z_{low}}^{z_{high}}h_{PSF}^{2}(w_{1}+x,z_{1}+y) \tag{40}$$

$$\sigma_{test}(x,y) = \sqrt{\sum_{w_{1}=w_{low}}^{w_{low}}\sum_{z_{1}=z_{low}}^{z_{high}}h_{PSF}^{2}(w_{1}+x,z_{1}+y)} \tag{41}$$

29

By using substitution the variance of the correlator, $\sigma_{corr}(x,y)$, can be fully defined as:

$$\sigma_{corr}(x,y) = \sigma_\delta(x,y)\sigma_{test}(x,y) \tag{30}$$

$$= \sqrt{\frac{\displaystyle\sum_{w=w_{low}}^{w_{high}}\sum_{z=z_{low}}^{z_{high}} d^2(w,z)}{N^2} - B(x,y)^2} \; \sqrt{\sum_{w=w_{low}}^{w_{high}}\sum_{z=z_{low}}^{z_{high}} h^2_{PSF}(w+x,z+y)} \tag{42}$$

Thus the both the point detector and the correlator are established as a zero-mean, unit-variance Gaussian random variable and can be compared to the same threshold on equal footing.

## 5. Search Methodology

The image of the satellite moves perceptibly from frame to frame. Therefore, the detectors are tried on all the pixels in a $6\times6$ search grid that is roughly centered on the satellite. The satellite is considered to be detected when it is found in any pixel in the search grid, which is the equivalent of applying a logical "or." When the satellite is detected in multiple pixels in a single frame, the SNR and probability of detection calculations shown later are based on the pixel in the frame that had the highest probability of detection.

For the purposes of this analysis, the eclipse is considered to last from the first non-saturated frame to the last non-saturated frame. The algorithm was only used to detect objects during this defined eclipse period for a two reasons. First, the objects are easily detectable in saturated frames by simply searching for pixels that have an intensity that

30

equals the maximum value of the ADC, so it is a trivial case. Second, at saturation charge spillover into nearby pixels distorts the data and saturation is in itself a significant non-linearity, both of which challenge the assumption that the system as a whole is essentially linear.

## IV. Results

### 1. Detections

The correlator showed an appreciable gain in the number of detections over the region in which the two detectors cross the threshold. Even though the experiment successfully showed the satellite pass through a large range of illumination, the time in which it passed through the threshold was relatively small compared to the size of the data set, and so there are a limited number of frames in which a two detectors gave a different answer. The output of each detector as the satellite goes into eclipse is plotted in Figure 10. The correlator did not universally perform better, but an examination of the detections near the threshold shows that the correlator generally detected the satellite deeper into the eclipse. Though instructive, simply noticing the additional frames of detection does not convey much knowledge about the relative performance of the detectors. The small number of frames during which the detectors straddle opposite sides of the threshold makes it difficult to draw well-supported, quantifiable conclusions.

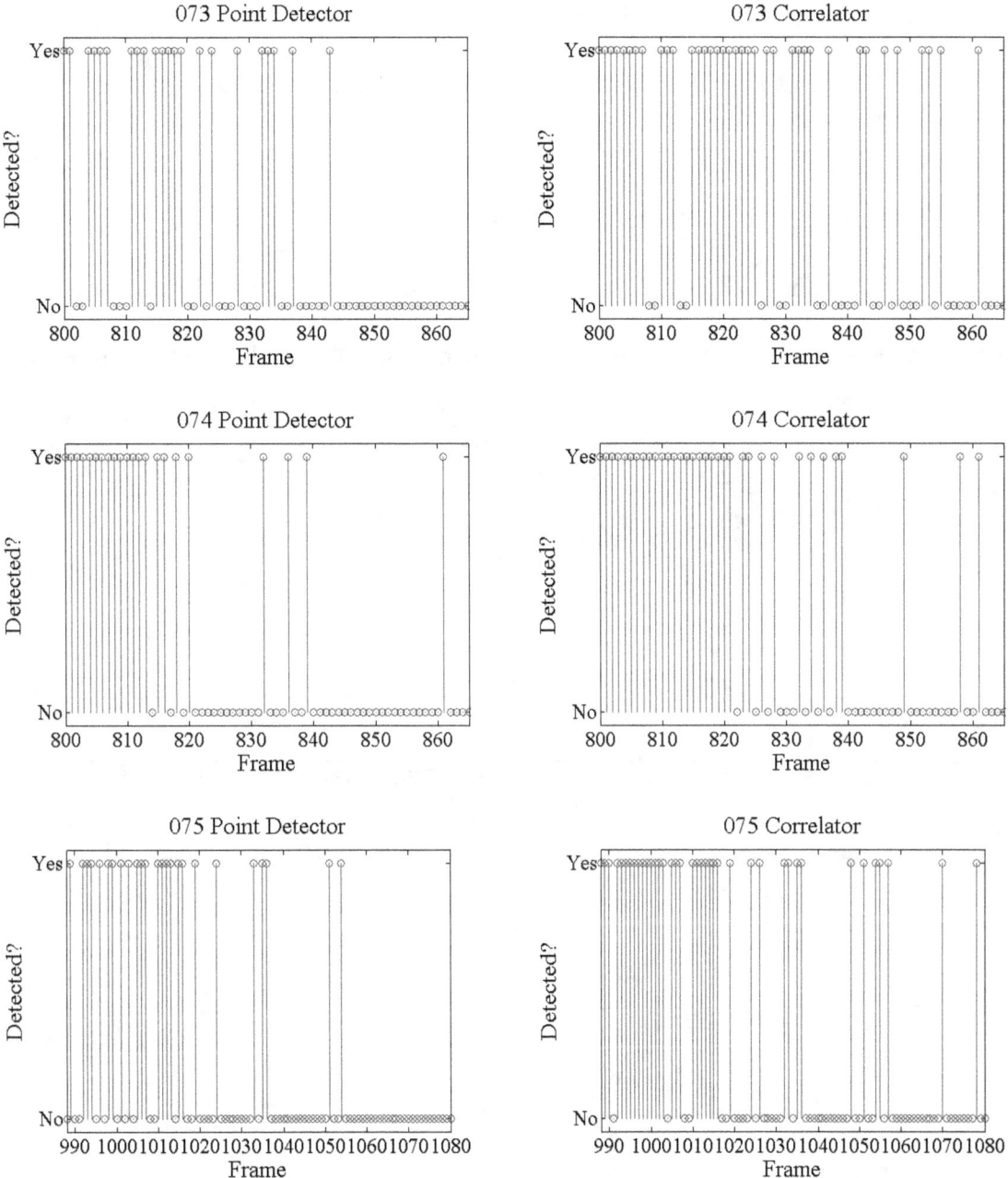

**Figure 10.** Detections as the satellite goes into eclipse for all three nights for both detectors near the crossing of SNR 6. The left and right columns show the results of the point detector and correlator, respectively. From top to bottom the three nights of data are 073, 074, and 075. The number of additional frames that the correlator was able to detect the satellite in may seem insignificant, but the span of frames shown represents a significant drop in apparent magnitude.

## 2. Comparison of Signal to Noise Ratio

As has been established in the discussion of the threshold, the output of the detectors for a given pixel directly corresponds with the signal to noise ratio (SNR) for that pixel. The algorithms register a detection when the SNR of a pixel rises above the threshold. Figure 11 shows the SNR of both detectors as the satellite goes into eclipse on night 073. The point detector can occasionally be seen to exhibit a higher SNR than the correlator,

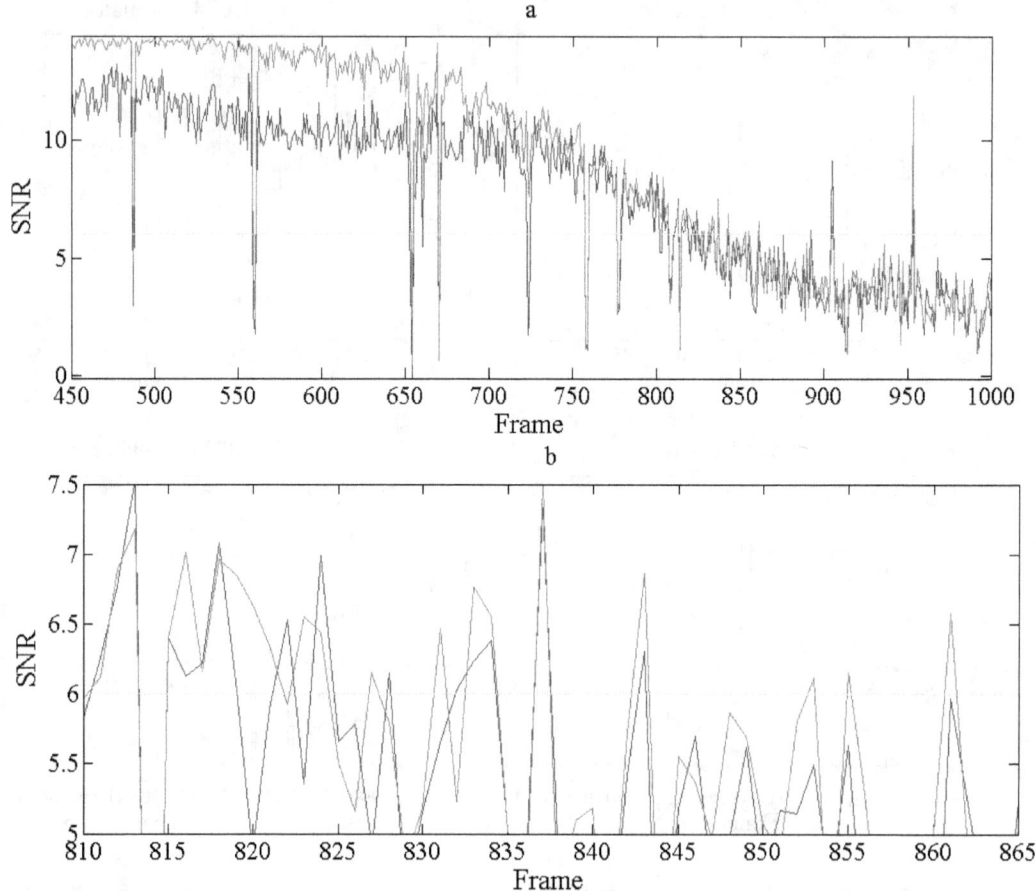

**Figure 11.** Signal to noise ratio for both detectors as the satellite goes into eclipse on night 073. The correlator is depicted in red and the point detector is depicted in blue. The horizontal green line shows the threshold of SNR 6. The top graph (a) shows the full data set and the bottom graph (b) shows a close up as the satellite fades beneath the threshold of detection. The sharp spikes in (b) around frames 900 and 950 are due to star crossings.

33

but, on average, the correlator comes out on top. The correlator also seems to experience less perturbation when other sources of light interfere.

The discussion that now follows concentrates on the region of the datasets where the output of the detectors is near the threshold, since those data points are the ones that allow for differentiation of the detector's performance. The reason is that objects that are much brighter are not interesting since they are easily detectable by both algorithms, and objects that are much dimmer are not interesting since the probability of detection by either algorithm approaches zero.

### 3. Relative Probability of Detection

Attempting to compute the probability of the detection directly poses several major hurdles. Most significantly, it requires foreknowledge of the presence of an object that is difficult to detect. The light-gathering power of the Space Surveillance Telescope (SST) makes it possible to see objects that have not previously been detected. Were this research conducted using images collected by a small aperture telescope, it might be possible to rely on a catalog to indicate where stars are that would appear dim to that telescope; this is obviously not a possibility for SST.

Similarly, attempting to compute the probability of detection directly is also extremely difficult. To do so requires absolute knowledge that there is absolutely nothing in a given window: a philosophical conundrum. Instead a threshold of detection is chosen in a way that determines the probability of false alarm based on assumptions in the Gaussian model.

None of the differences between the point detector or correlator change the probability of false alarm. Holding this variable of performance constant allows for the algorithms to be compared with relative ease. While the absolute probability of detection cannot easily be determined for either method, it is possible to determine the probability of detection of one as a function of the probability of detection of the other using an analysis of the signal to noise ratios (SNRs).

The starting frame for the graphs that will be shown in this section is when probability of detection for the two detectors starts to recede from one, and the ending frame is when the probability of detection for the two detectors approaches zero. Including points outside of this range results in clutter near the endpoints that obscures the features of the graphs and does not add any meaningful information.

For a given object with a specific SNR, the probability density function for the probability of detection is predicted by a unit-variance ($\sigma$), Gaussian distribution, $f(x; \mu, \sigma^2)$:

$$f(x; \mu, \sigma^2) = \frac{1}{\sigma \sqrt{2\pi}} e^{-\frac{1}{2}\left(\frac{x-\mu}{\sigma}\right)^2} \tag{43}$$

Since the Gaussian distribution is symmetric, when the measured SNR, $x$, is set to equal the mean, $\mu$, the probability density reaches a maximum value.

Obviously, the likelihood of detection increases as the threshold SNR gets lower, just as the probability of false alarm also increases. In the same way, the probability of detection, $Pd$, decreases as the threshold SNR is raised. Thus, the PDF should be integrated beginning from a threshold SNR of positive infinity, towards a lower bound threshold, $\tau$, as in:

$$Pd = \int_{\tau}^{\infty} f(x;\mu,\sigma^2)dx \qquad (44)$$

Checking the upper and lower bounds for the threshold shows that this is sensible. When the threshold of detection is set impossibly high, such that $\tau = \infty$, we see that

$$Pd = \int_{\tau=\infty}^{\infty} f(x;\mu,\sigma^2)dx = 0 \qquad (45)$$

If $\tau = -\infty$ then the SNR is set as low as possible so that there is no requirement for any signal at all; in this case the integration is defined as one:

$$Pd = \int_{\tau=-\infty}^{\infty} f(x;\mu,\sigma^2)dx = 1 \qquad (46)$$

Taking notice that the above integration are not from "left to right," the notation can now be simplified by using the definition of the cumulative distribution function (CDF) of the Gaussian distribution, $F(\tau;\mu,\sigma^2)$, and subtracting it from one [12]:

$$Pd = 1 - F(\tau;\mu,\sigma^2) \qquad (47)$$

Recall that the mean of a detection algorithm is the threshold at which an object with a specific SNR is equally likely to be detected as not. The relative probability of detection of the two algorithms can then be compared for the same threshold, $\tau$, since the sufficient statistics of the two algorithms are unit variance Gaussian random variables and they are being compared in detecting the same object; the only difference is the mean of the distributions.

Consider a single data point in which the threshold is SNR five and the mean of the correlator is SNR six for an object with SNR six. The relative probability of detection for the point detector is:

$$Pd_\delta = 1 - F(\tau; \mu_\delta, \sigma^2) = 1 - F(6,5,1) \approx 1 - 0.83 = 0.17 \qquad (48)$$

Likewise, the relative probability of detection of the correlator is:

$$Pd_{corr} = 1 - F(\tau; \mu_{corr}, \sigma^2) = 1 - F(6,6,1) = 1 - 0.5 = 0.5 \qquad (49)$$

Each frame presents both algorithms with the same snapshot of an object. To simplify the computations, the threshold of comparison is chosen as six. As the Gaussian distribution swells at the mean, it is assumed that the mean of each algorithm is the SNR at which the object was detected.

Having covered the general theory, the notation is now expanded to include an index to denote the frame from which the SNR is derived and for which the probability of detection is being computed. Then, for frame £ the probability of detection for each detector is

$$Pd_\delta(f) = 1 - F(x(f); \mu_\delta(f), \sigma^2) = 1 - F(6, SNR_\delta(f), 1) \qquad (50)$$

and

$$Pd_{corr}(f) = 1 - F(x(f); \mu_{corr}(f), \sigma^2) = 1 - F(6, SNR_{corr}(f), 1) \qquad (51)$$

This comparison only yields a relative comparison of the probability of detection of each algorithm. Since it depends on the threshold of detection being the same for both algorithms, and since both algorithms typically detect objects that are much brighter than the threshold, it gives an estimate of the relative increase of the objects that will be detected around SNR six, not an increase in the objects that will be detected in total. This

37

is reasonable, since we would not expect to suddenly be able to detect a larger number of relatively bright objects by changing to a different algorithm.

Since integration of the Gaussian curve is very sensitive to the placement of the mean, computing the probability of detection in this manner across a region of the dataset yields a very noisy signal. Implementing a low-pass filter on the SNR signal that before it is fed into the integration removes the high-frequency noise and gives a better picture of the performance trends of the two detectors. The low-passed signal, $\overline{SNR}(f)$, is generated by taking the running average of 15 frames of data[13]:

$$\overline{SNR}(f) = \frac{\sum_{g=1}^{15} \overline{SNR}(g)}{15} \tag{52}$$

The affect of this low pass filter on the SNR is shown in Figure 12, which shows the instantaneous (top) and low-passed (bottom) SNR's for both detectors as the satellite goes into eclipse on night 073.

Low-passed versions of the probability of detection are then defined based on the low-passed $\overline{SNR}(f)$:

$$\overline{Pd_\delta}(f) = 1 - F(6, \overline{SNR_\delta}(f), 1) \tag{53}$$

$$\overline{Pd_{corr}}(f) = 1 - F(6, \overline{SNR_{corr}}(f), 1) \tag{54}$$

Possibly the most illuminating comparison is when the probability of detection for each of the algorithms is compared per frame. Figure 13 shows plots of $Pd_\delta$ and $Pd_{corr}$ over a span of adjacent frames as the probability of detection sweeps from near zero, through the threshold of detection, to practical certainty for the two algorithms as the

38

satellite goes into and out of eclipse. Also shown is a comparison of the low-passed $\overline{Pd_\delta}$ and $\overline{Pd_{corr}}$, which give a clearer picture of the generalized performance of the correlator on a given eclipse event. Since the point detector is currently in use at LINEAR and SST it is shown as the baseline for performance.

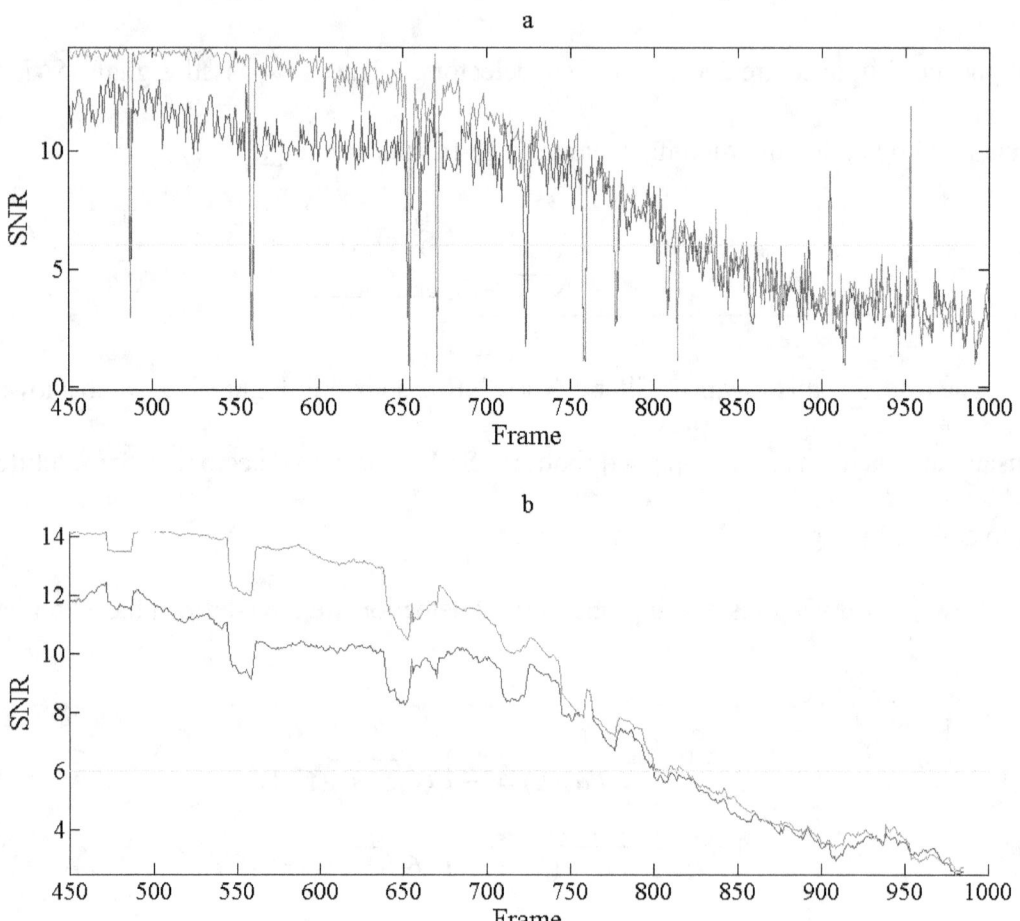

**Figure 12.** Instantaneous (a) and low-passed (b) SNR of the detectors as the satellite goes into eclipse on night 073. The correlator is depicted in red and the point detector is depicted in blue. The horizontal green line shows the threshold of SNR 6. Note how the low-passed SNR reveals an undulating pattern of lower sensitivity that affects both detectors.

39

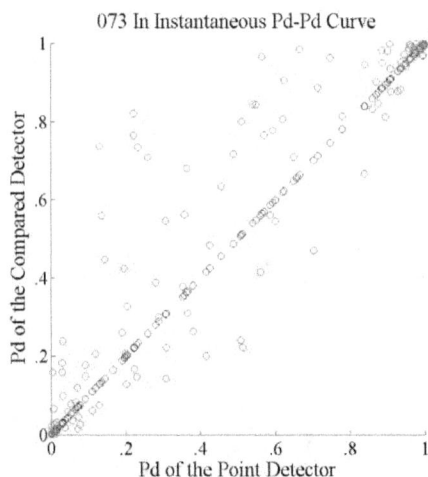

073 In Instantaneous Pd-Pd Curve

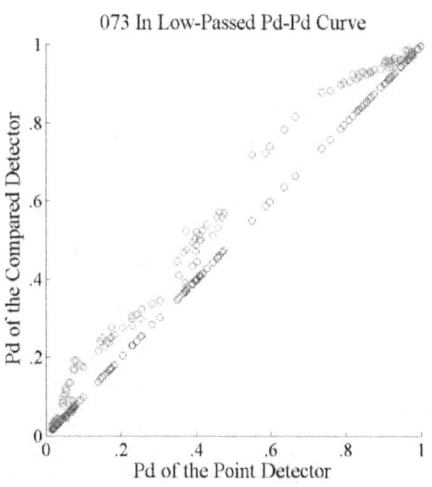

073 In Low-Passed Pd-Pd Curve

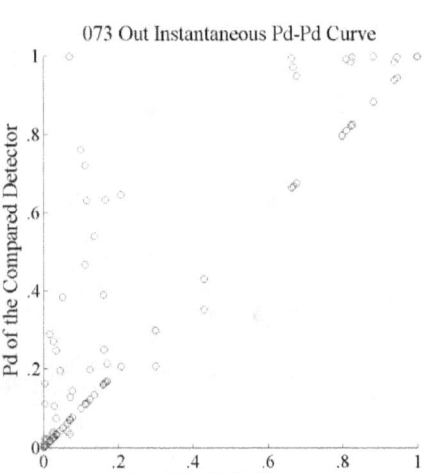

073 Out Instantaneous Pd-Pd Curve

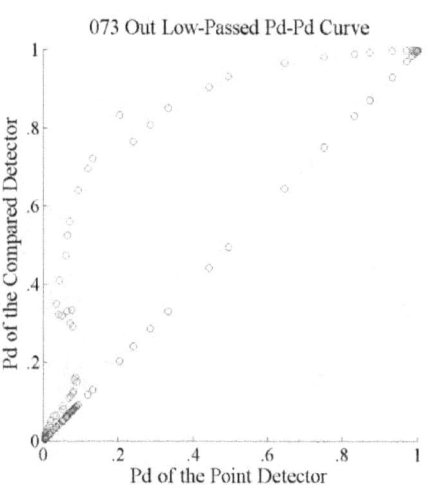

073 Out Low-Passed Pd-Pd Curve

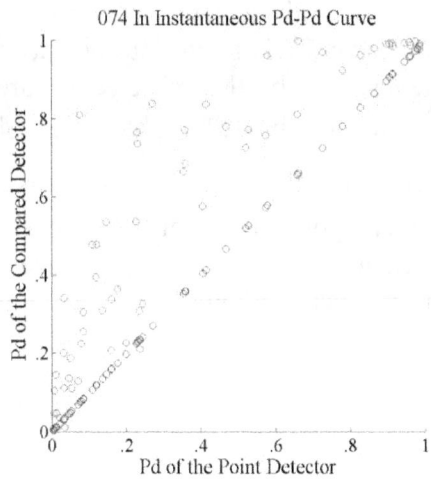

074 In Instantaneous Pd-Pd Curve

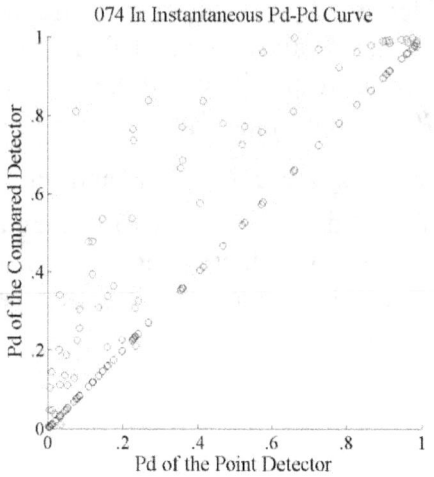

074 In Instantaneous Pd-Pd Curve

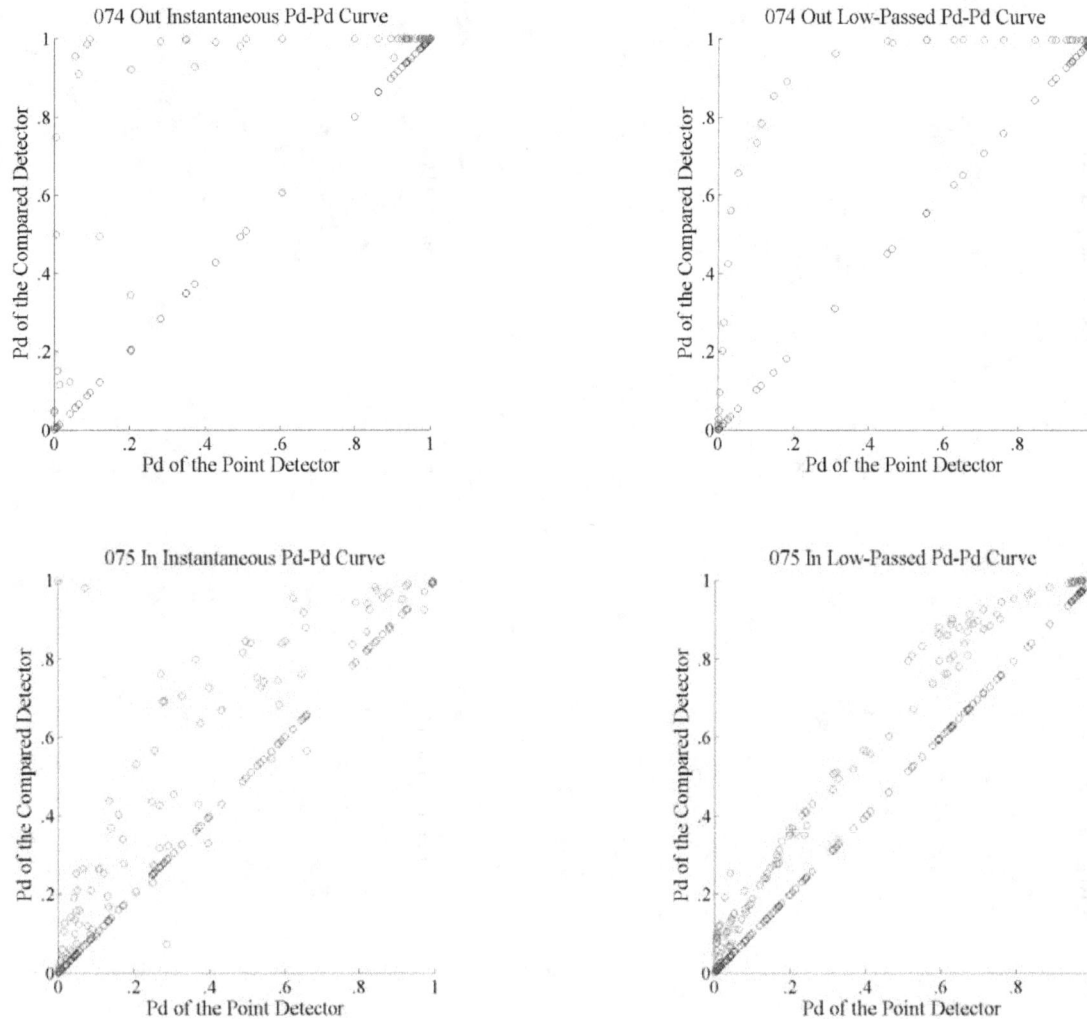

**Figure 13.** Comparison of the probability of detection for both detectors on all three nights. Graphs are labeled "In" or "Out" for whether the satellite is going into or out of eclipse, respectively. In the left column the instantaneous probabilities of detection are $Pd_\delta$ in blue and $Pd_{corr}$ in red. In the right column the low-passed probabilities of detection are $\overline{Pd_\delta}$ in blue and $\overline{Pd_{corr}}$ in red, both of which are the product of low-pass-filtered data that uses a 15 frame running average. The data points used for these graphs are those which are near the threshold of detection for each of the algorithms; they were arranged in order of increasing probability of detection when computed by the point detector.

41

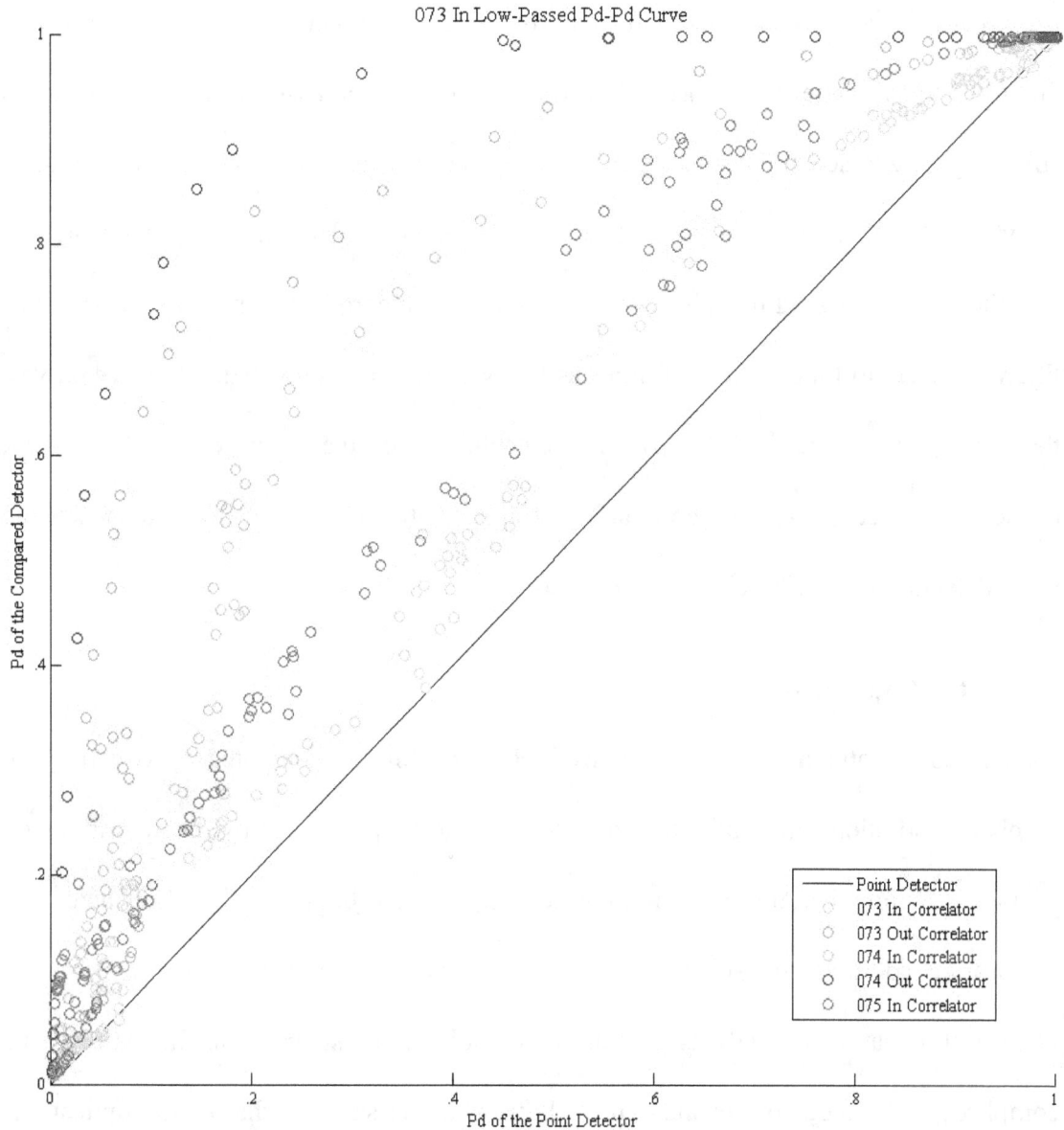

**Figure 14.** Co-plotted curves of $\overline{Pd_{corr}}$ for all five recorded eclipse events over nights 073, 074, & 075.

A condensed view of the relative probability of detection of both detectors over all five eclipse events over the three nights is shown in Figure 14. The differences in the low-passed curves shows that the performance gains of the correlator vary significantly from a small, though still meaningful, improvement as the satellite enters the eclipse on night 073 to over a near-guaranteed detection for the correlator while the point detector is still struggling to detect the satellite as it exits eclipse on the night of 074. Over the range in which the point detector has between a 20% and a 50% chance of detecting the satellite, the low-passed probability of detection for the correlator ranges between just as likely to detect to four-and-a-half times as likely to detect. Generalizing the five curves, the correlator averages a 75% chance of detection when the point detector has a 50% chance of detection, and the correlator averages a 50% chance of detection when the point detection has a 20% chance of detection.

## 4. Cost to Implement

The computational burden of each of the algorithms is computed by counting the number of additions and multiplications that have to be performed by the algorithms on average. While certain programming techniques can be result in efficiencies for performing addition and subtraction, modern superscalar processors perform floating point multiplication as easily as floating point addition and subtraction. Incredibly, "the complexity of square root evaluation or division is the same as that of multiplication" when the Newton-Raphson method is used [14]. Thus floating point operations for addition, subtraction, multiplication, division, and square roots are assumed to impose the same computational burden. Furthermore, it is assumed that element-wise multiplications of matrices are performed using a matrix or vector processing unit in

43

which loading the appropriate data based on a region does not impose an appreciable computational burden.

First, consider the formula for the point detector, $\mathsf{L}_\delta(x,y)$:

$$\mathsf{L}_\delta(x,y) = \frac{(d(x,y) - B(x,y))}{\sigma_\delta(x,y)} \qquad (22)$$

where $(x,y)$ is the pixel being examined, $d(x,y)$ is the data point in that pixel, $B(x,y)$ is the background in an $N \times N$ window, and $\sigma_\delta(x,y)$ is the standard deviation of the data in the window. Given $B(x,y)$ and $\sigma_\delta(x,y)$, evaluating the point detector requires only a subtraction and division, which is two operations.

The background, $B(x,y)$, is approximated for each window by taking the mean:

$$B(x,y) = \frac{\displaystyle\sum_{w=w_{low}}^{w_{high}} \sum_{z=z_{low}}^{z_{high}} d(w,z)}{N^2} \qquad (18)$$

Evaluating the numerator requires $N^2 - 1$ additions, evaluating the denominator requires a single multiplication, and the taking the ratio requires a single division. As $N$ increases, the number of operations needed in determining it rapidly converges to

$$\left(N^2 - 1\right) + 1 + 1 \approx N^2 \qquad (55)$$

Likewise, $\sigma$, the standard deviation of the background, is given by:

$$\sigma_\delta(x,y) = \sqrt{\frac{\displaystyle\sum_{w=w_{low}}^{w_{high}} \sum_{z=z_{low}}^{z_{high}} d^2(w,z)}{N^2} - B(x,y)^2} \qquad (24)$$

44

Squaring $d(w,z)$ for each pixel requires $N^2$ multiplications, and the summation itself requires $N^2 - 1$ additions. Dividing by $N^2$ adds only two operations, as with evaluating the background. Subtracting the background requires only a single operation since it has already been computed, and one more again for the square root. As $N$ increases, the number of operations needed in determining $\sigma$ rapidly converges to

$$N^2 + \left(N^2 - 1\right) + 2 + 1 + 1 \approx 2N^2 \tag{56}$$

Recalling the definition of the point detector, altogether evaluating the point detector requires $N^2$ for the background, $2N^2$ for the standard deviation, and an additional subtraction and multiplication, so the total number of operations needed converges to

$$N^2 + 2N^2 + 2 \approx 3N^2 \tag{57}$$

Second, consider the formula for the correlator:

$$\mathsf{L}_{corr}(x,y) = \sum_{w=w_{low}}^{w_{high}} \sum_{z=z_{low}}^{z_{high}} \frac{(d(w,z) - B(x,y))h_{PSF}(w+x, z+y)}{\sigma_{corr}(x,y)} \tag{28}$$

where $(x,y)$ is the center pixel of the $N \times N$ window being correlated, $w$ and $z$ are indexes in the window, $d(w,z)$ is the data point at a specific index, $B(x,y)$ is the background in the window, $h_{PSF}(w+x, z+y)$ refers to the pixel of the PSF for a point source of the system that is centered at $(x,y)$ and has been normalized for the window, and $\sigma_{corr}(x,y)$ is the standard deviation of the window.

From Equation (30), the standard deviation for the correlator is found by multiplying the standard deviation for the point detector by $\sigma_{test}(x,y)$, which is defined as:

45

$$\sigma_{test}(x,y) = \sqrt{\sum_{w_1=w_{low}}^{w_{low}} \sum_{z_1=z_{low}}^{z_{high}} h^2_{PSF}(w_1 + x, z_1 + y)} \qquad (41)$$

Squaring $h^2_{PSF}(w_1 + x, z_1 + y)$ element-wise and taking the summations, followed by the square root, requires $2N$ operations. However, since Equation (41) is not dependent on $d(w,z)$ or any terms that are dependent on $d(w,z)$, it does not change as $(x,y)$ changes; thus, it can be computed once for the entire data set. As a result, there is not an appreciable computational cost in calculating $\sigma_{test}(x,y)$.

Expanding Equation (28) and pulling the division out of the summation yields

$$L_{corr}(x,y) = \frac{\sum_{w=w_{low}}^{w_{high}} \sum_{z=z_{low}}^{z_{high}} d(w,z)h_{PSF}(w+x,z+y)}{\sigma_{corr}(x,y)} - B(x,y)\frac{\sum_{w=w_{low}}^{w_{high}} \sum_{z=z_{low}}^{z_{high}} h_{PSF}(w+x,z+y)}{\sigma_{corr}(x,y)} \qquad (58)$$

Since the PSF has been normalized over the window, the summation of the PSF over the window is identically one, so Equation (58) can be simplified to:

$$L_{corr}(x,y) = \frac{\sum_{w=w_{low}}^{w_{high}} \sum_{z=z_{low}}^{z_{high}} d(w,z)h_{PSF}(w+x,z+y)}{\sigma_{corr}(x,y)} - \frac{B(x,y)}{\sigma_{corr}(x,y)} \qquad (59)$$

Since there are $N^2$ pixels in the window, one operation on a single pixel of the summation requires $N^2$ operations, so multiplying $d(w,z)$ and $h_{PSF}(w+x,z+y)$ and then performing the summation requires two operations per pixel, or $2N^2 - 1$ operations in total. Dividing the summation by the standard deviation and then subtracting the ratio of the background to the standard deviation requires only three more operations. Thus, the additional computational cost of correlator over the point detector converges to

46

$$(2N^2 - 1) + 3 \approx 2N^2 \qquad (60)$$

Recalling that the cost of implementing the point detector is $3N^2$, the cost of the correlator is

$$3N^2 + 2N^2 = 5N^2 \qquad (61)$$

Thus, the correlator requires 67% more operations than the point detector when $N$ is chosen in accordance with current practice [15].

## V.  Conclusions

### 1.  The Correlator Performs Better

The correlator outperforms the point detector in terms of signal to noise ratio (SNR) in nearly all frames.  More importantly, the correlator yields a higher probability of detection of dim objects near the threshold of detection.  Based on the five low-passed probability of detection curves, the correlator averages a 75% chance of detection when the point detector has a 50% chance of detection, and the correlator averages a 50% chance of detection when the point detection has a 20% chance of detection.  Thus, it is possible that, without changes made to the way data is collected and without an increase in the likelihood of false alarm, dim objects will be detected more frequently and even dimmer objects will be detectable.

### 2.  Minimal Cost to Implement

Very few changes are necessary in implementing the correlator versus the point detector.  The correlator only requires 67% more calculations than the point detector.

47

Additionally, the additional calculations do not extend the depth of the pipeline required, and so the additional computational cost can easily be accommodated by adding hardware in parallel.

## 3. Limitations and Possible Future Work

The output of the correlator changes less than the output of the point detector when other sources of illumination are in the window. Should a particular task require this sensitivity, some adjustments may need to be made. This also suggests that the correlator may be better able to differentiate between closely spaced objects. The detection of dim objects in the presence of other objects, especially brighter objects, could be a very fruitful area of study.

Both methods still rely on a Gaussian model of the emission or reflection of photons from the dim object, which is inaccurate. In truth, photon arrival is best described by the Poisson distribution. The noise of the amplifiers in the circuit does create Gaussian noise, but any noise that results in a voltage lower than signal-ground is registered by the analog-digital converters (ADCs) as a zero. The Poisson-based log-likelihood ratio test detector that Capt Peterson developed could be adapted to collections from the SST that lacks a known background, but establishing a metric for comparing Poisson- and Gaussian-based detectors remains a challenge.

The method of selecting a PSF used in this research is labor intensive and highly subjective. Furthermore, since the imaging plane is under-sampled for the typical seeing conditions, it is likely that the performance of the correlator suffers from the affects of aliasing. The appearance of structure in the periodic dips of the SNR of the detectors supports the theory that aliasing is happening as small shifts of the satellite on the imaging plane take place.

## 4. Summary

The correlator offers significant performance advantages for the LINEAR and SST programs for little to no cost depending on the available computing resources. In particular, the correlator is able to capitalize on the unprecedented light-gathering ability of SST and enable it to see hither-unknown or undiscoverable dim objects.

# Bibliography

[1]     Congress, United States 109th. *Public Law 109–155*. s.l.: United States Government, 2005. United States Legislation.

[2]     DARPA. Space Surveillance Telescope. [Online]. [Cited: 01 17, 2013.] http://www.darpa.mil/Our_Work/TTO/Programs/Space_Surveillance_Telescope _(SST).aspx

[3]     DARPA. SST Fact Sheet. [Online]. [Cited: 01 17, 2013.] http://www.darpa.mil/uploadedImages/Content/Our_Work/TTO/Programs/SST/ FactSheetTabFINAL.JPG

[4]     Grant H. Stokes, F. C. Shelly, M. S. Blythe, and J. S. Stuart. "Applying Electro-Optical Space Surveillance Technology to Asteroid Search and Detection: The Linear Program Results." *Proc. 1998 Space Control Conf.* Lexington, 1998.

[5]     Stephen C. Pohlig. "An Algorithm for Detection of Moving Optical Targets." *IEEE Transactions on Aerospace and Electronic Systems*. Vol. AES-25, no. 1 (1989, Jan.): 56-63.

[6]     Joseph W. Goodman. *Statistical Optics*. New York: John Wiley & Sons, 1985.

[7]     Curtis J. R. Peterson. "Near Earth Object Detection Using a Poisson Statistical Model for Detection on Images Modeled from the Panoramic Survey Telescope and Rapid Response System," M.S. thesis, E.E. Dept., Air Force Inst. of Tech., Wright Patterson AFB, OH, 2012.

[8]     Joseph W. Goodman. *Introduction to Fourier Optics*. Greenwood Village, CO: Roberts & Company Publishers, 2005.

[9]     Michael Vollmer and Stanley David Gedzelman. "Simulating Irradiance During Lunar Eclipses: the Spherically Symmetric Case." *Applied Optics*. Vol. 47, no. 34 (2008, Dec.).

[10]    Frederick J. Vrba, et al. "A Survey of Geosynchronous Satellite Glints." *Proc. 2009 AMOS Technical Conf.* Maui, 2009.

[11]    Richard D. Richmond and Stephen C. Cain. *Direct-Detection LADAR Systems*. Bellingham, WA: SPIE Press, 2010.


[12]    George Casella and Roger L. Berger. *Statistical Inference*. Crawfordsville, IN: R. R. Donnelley & Sons, Co., 2002.


[13]    Emmanuel C. Ifeachor and Barrie W. Jervis. *Digital Signal Processing: A Practical Approach*, 2nd, Ed. New York: Prentice Hall, 2002.


[14]    Jean-Michel Muller. *Elementary Functions: Algorithms and Implementation*. Boston: Birkhäuser, 2006.


[15]    Stephen Maksim, John Zingarella, and Stephen Cain. "A Comparison Between a Non-linear, Poisson-based Statistical Detector and a Linear, Gaussian Statistical Detector for Detection Dim Satellites." *Advanced Maui Optical and Space Surveillance Technologies Conf.* Maui, 2012.